Alan McKirdy has written many popular books and book chapters on geology and related topics and has helped to promote the study of environmental geology in Scotland. Before his recent retirement he was Head of Knowledge and Information Management at Scottish Natural Heritage. He is now a freelance writer.

Skye

LANDSCAPES IN STONE

Alan McKirdy

For Nancy and Peter

First published in Great Britain in 2016 by
Birlinn Ltd
West Newington House
10 Newington Road
Edinburgh
EH9 1QS

www.birlinn.co.uk

ISBN: 978 1 78027 372 3

Copyright © Alan McKirdy 2016

The right of Alan McKirdy to be identified as the author of this work has been asserted by him in accordance with the Copyright, Designs and Patents Act, 1988

All rights reserved. No part of this publication may be reproduced, stored, or transmitted in any form, or by any means, electronic, mechanical or photocopying, recording or otherwise, without the express written permission of the publisher.

British Library Cataloguing-in-Publication Data
A catalogue record for this book is available on request from the British Library

Designed and typeset by Mark Blackadder

FRONTISPIECE.
Neist Point, Skye

Printed and bound by Latimer Trend, Plymouth

Contents

	Introduction	7
	Skye through time	8
	Geological map	10
1.	It's a dynamic Earth	11
2.	From the beginning	13
3.	Dinosaur island	16
4.	The Skye volcano	25
5.	The Ice Age	34
6.	Dynamic landscapes	36
7.	Living landscapes	38
8.	Places to visit	42
	Acknowledgements and Picture Credits	48

Introduction

The Isle of Skye offers the resident and visitor alike a magical combination of wild land and breathtaking natural beauty that makes this place unique in the British Isles. This book provides a starting point for an understanding and appreciation of how the geology and landscapes of the island came to be. In addition, the ongoing processes of natural landscape evolution that continue to leave their mark on these spectacular vistas are also described.

Skye's geological history is a long and tortured affair, involving some of the most ancient rocks on the planet; a grandstand view as the Highlands of Scotland were formed over 400 million years ago and the development of one of the mightiest volcanoes ever to blow its top in geological history. Skye is also known as Scotland's 'dinosaur island' yielding the remains of plant- and meat-eating creatures that stalked the Jurassic landscapes some 170 million years ago. And finally, the rocks forged in earlier times were shaped and fashioned into the familiar hills and glens of today by the passage of ice as a great freeze gripped the land. It's quite a story!

In this book, we draw on the many scientific books and academic papers that describe this beautiful and geologically diverse place and trust it whets the appetite of all who want to understand something of the scenic grandeur of this amazing place.

Opposite. Glen Brittle, view of the Black Cuillin.

Skye through time

Period of geological time	Millions of years ago	Scotland's global position	Environments and events on Skye
Anthropogene	Last 10,000 years	57° N	This is the age of *Homo sapiens* – us! After the ice melted, we colonised the land and made the island habitable. Trees were felled and the last wolf, our only competition as top predator, was shot in 1745. By our actions, we continue to modify the landscape, atmosphere and climate to this day.
Quaternary	Started 2 million years ago	Present position of 57° N	This is the age of ice. Many advances and retreats of the ice are recorded during this period. The landscapes of Skye were created by glaciers during these times. The solidified magma chamber of the now silent Skye volcano was revealed as the ice bit deep into its heart.
Neogene	2–24	55° N	The climate cools as the Ice Age approaches.
Palaeogene	24–65	50° N	Europe split apart from North America as the Atlantic Ocean widened and the crust became stretched. The Skye volcano was active at this time (from 60 million to 55 million years ago), with great volumes of basalt lava erupted onto the surface and a reservoir of molten rock developed at depth.
Cretaceous	65–142	40° N	The land that was to become Skye was submerged under a warm tropical sea, as the world's oceans rose to cover much of Scotland.
Jurassic	142–205	35° N	Thick layers of sandstone and limestone were deposited under a shallow sea. These waters teemed with life, now preserved as fossils in the rock. Dinosaur remains and footprints have been recovered from rocks of this age, as well as ammonites and fossil plants.
Triassic	205–248	30° N	Sandstone and layers of rounded cobbles of this age are found around Broadford and also on Raasay.

Period of geological time	Millions of years ago	Scotland's global position	Environments and events on Skye
Permian	248–290	20° N	Scotland continued to drift northwards. No rocks of this age are recorded on Skye.
Carboniferous	290–354	On the Equator	No rocks of this age are recorded on Skye.
Devonian	354–417	10° S	Scotland was located 10°S of the Equator at this time. No rocks of this age are recorded on Skye.
Silurian	417–443	15° S	The Caledonian uplands were formed during this time as the mountain-building episode was completed.
Ordovician	443–495	20° S	The Iapetus Ocean reached its widest point. The land that was to become Skye lay at the northern margin of this ocean. Rocks from these times were beach deposits and other strata were laid down close to the shoreline. The continental collision that led to the formation of the Caledonian Mountains started in late Ordovician times as landmasses converged.
Cambrian	495–545	30° S	
Proterozoic	545–2,500	Close to South Pole	Torridonian sandstones were dumped on top of the Lewisian basement by fast-flowing rivers and streams. Earlier Moine rocks are of enigmatic origin.
Archaean	Prior to 2,500	Unknown	The northern part of Raasay and part of the Sleat Pennisula are built from ancient rocks, known as Lewisian gneisses. They are the most ancient of all, dating back some 2.8 billion years.

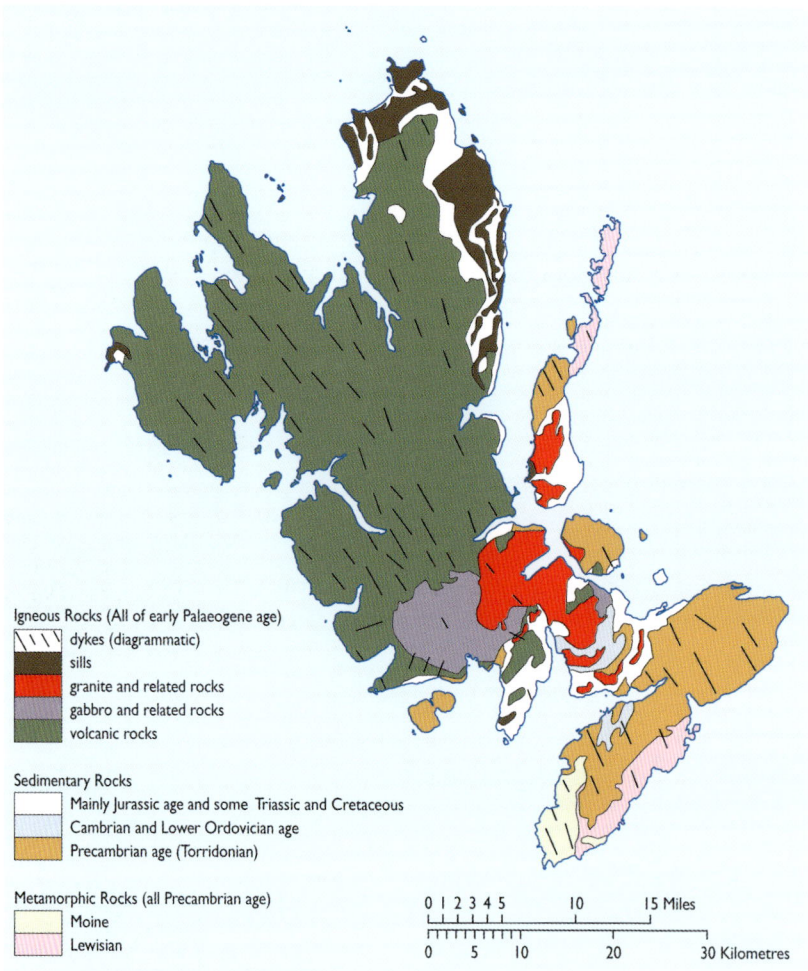

Geological map of Skye. The rocks of Skye are codified by their age and origin on this geological map. The greater part of Skye and its surrounding islands is built from rocks associated with the Skye volcano. The north and west of the island are made of lavas erupted from the volcano, whilst the Black and Red Cuillin are the ancient magma chambers that fed the volcano. The magma chambers were later excavated by ice in the more recent geological past. More ancient rocks (Lewisian, Moine and Torridonian) underlie much of the Sleat Peninsula, Soay, Raasay and South Rona. Limestones and shales from the Jurassic Period are a feature of the Trotternish Peninsula, Raasay and the areas around Elgol and Broadford. The most recent event, the glaciation of the island during the Ice Age, left a strong imprint on the scenery, but deposits and other features that date from this time (glacial moraines and erosional features for example) are not represented on this map.

1
It's a dynamic Earth

Even in this short geological history of Skye, it's important to appreciate how Planet Earth functions and how geological processes, past and present, have helped to shape the landscape we see today. We may think of the Skye Cuillin and the other great mountains of Scotland as immovable, immutable and unchanging since time began, but evidence from the world around us tells a different story.

The Earth's crust is divided into continent-sized chunks, called tectonic plates, which are constantly on the move. It's mainly at the edges of these plates where volcanoes, earthquakes and tsunamis are created.

At the plate boundaries, chunks of crust grind past each other, or collide with each other, or pull apart as new crust is created. The

Earth's tectonic plates. The plates are of uneven size but together cover the full extent of the Earth's surface.

From early beginnings close to the South Pole, the land that became Scotland has moved steadily northwards to its present position, 57° N. On this journey, it has passed through all of the Earth's climatic zones and collided with many other continents on the way.

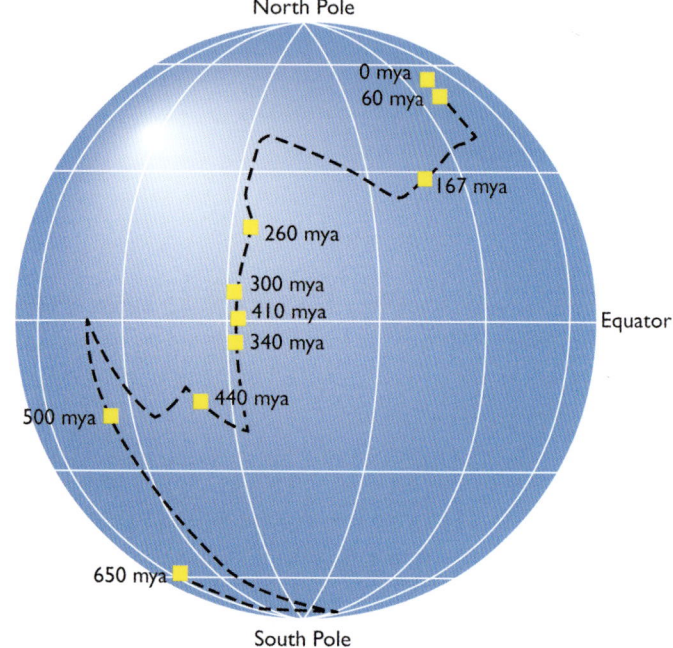

driving force for this movement emanates from the Earth's core, which at a colossal temperature of 6,000°C, heats up the overlying mantle and sets up a convective motion that carries the tectonic plate across the globe at a rate of around 6 centimetres each year. This may not seem particularly fast but, over millions of years, each and every one of the world's land areas has been propelled huge distances across the Earth's surface.

Mountain ranges, such as the Scottish Highlands and the Himalayas, are formed when two continental plates find themselves on a collision course. The sands, muds, limestones and lavas that accumulated in the ocean deep and coastal fringes are crushed, folded, buckled and uplifted into mountain ranges as a continent-to-continent 'head-on car crash' takes place. Many of the upland areas across the globe are the result of these continental collisions.

The presence of ancient and now extinct volcanoes is another potential source of spectacular scenery, and the story of the Skye volcano, which will be told later, is the origin of most of the higher ground of Skye.

2
From the beginning

The oldest rocks on Skye are found on the Sleat Peninsula, Raasay and South Rona. These Lewisian gneisses are also some of the oldest rocks to be found anywhere in Europe. They were formed around 2,800 million years ago from a wide variety of even older rocks that were modified by the effects of heat and intense pressure deep in the Earth's crust. The ancient gneisses are characteristically streaked and banded, and any remaining traces of the original rock types have been obliterated by repeated earth movements during their long history.

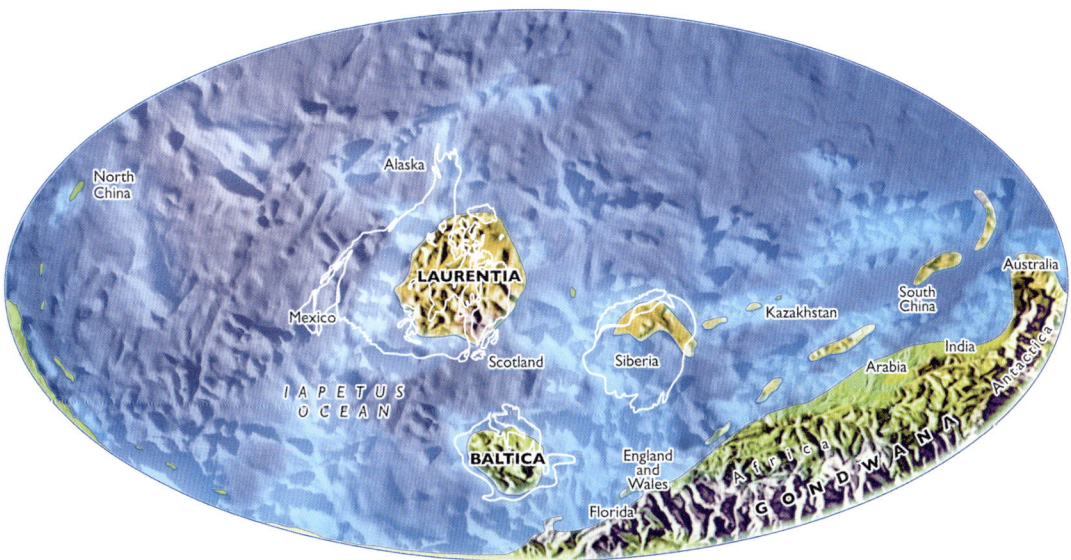

Planet Earth was a very different place 600 million years ago. The geography of the world from these times is unrecognisable today. Continental alliances existed that have since been torn asunder, and great oceans occupied much of the Earth's surface. The Iapetus Ocean existed for around 250 million years and played a huge role in the creation of 'Scotland'. Great thicknesses of sand and mud built up on the ocean floor during this time. As the Iapetus Ocean closed and Laurentia collided with the land we now recognise as England and Wales, these sediments were squashed and turned into rock. The product of this mountain-building process is the Highlands of Scotland, albeit much altered subsequently by effects of erosion by wind, water and ice.

By about 1,100 million years ago, these Lewisian rocks had been raised to the Earth's surface by powerful earth movements and had subsequently been worn away by wind and water to form a hummocky landscape. This surface was then gradually buried beneath several thousand metres of gritty and pebbly sandstones, known as Torridonian sandstone – the products of fast-flowing rivers that swept across an otherwise hot, dry landscape of the long-disappeared continent of Laurentia. The diagram on page 13 explains the ancient continent of Laurentia's place in the scheme of things.

At this time, the Lewisian gneisses and their Torridonian cover were part of a huge continent known as Laurentia, the remains of which now form large parts of Canada and Greenland. North-west Scotland lay on the south-eastern margin of this continent, the coastline of which was lapped by the waters of the Iapetus Ocean. At the edge of these primordial seas, sands, silts and limy muds were laid down during the Cambrian and Ordovician Periods. These sediments became rock and are recognised today as Durness limestones and associated quartzites. Some of the early signs of life have been recovered from these rock strata.

This tranquillity was shattered as the Iapetus Ocean began to close. As continents collided, the vice-like grip created as the landmasses approached each other caused intense folding, alteration by heat and

This is evidence for the existence of a worm that created tunnels and burrows in the sands that built up along the coastline of the Iapetus Ocean. These trumpet-shaped burrows gave rise to a rock known as 'pipe rock'. They are some of the earliest signs of life recorded in Scotland's geological record.

pressure, and even melting. The metamorphic and igneous rocks created by this collision process were lifted up as the Iapetus Ocean closed to form the Caledonian Mountains, the roots of which now constitute the bulk of the mainland Scottish Highlands. In scale, this was probably similar to the present-day Alps. The north-western limit of these effects passes through the Sleat Peninsula, where the Lewisian gneisses, the Torridonian sandstones and a group of metamorphosed sandstones called the Moine have all been sliced up and folded by the earth movements.

These rocks were completely mangled by movements in the Earth's crust during Skye's early geological history.

3
Dinosaur island

Over the next 200 million years, 'Scotland' continued to chart a course that took this chunk of the Earth's crust out of the southern hemisphere and northwards beyond the Equator. But there is, amazingly, little record of this span of time in the rocks of Skye.

The story is picked up again in Triassic times some 248 million years ago. By this time, Skye was part of a hot, dry desert area in which silts, sands and gravel were deposited periodically on wide floodplains. Thick layers of sands and small cobblestones piled up as rivers flowed

All the Earth's landmasses had converged at this point to create a super-continent known as Pangaea (from the Greek meaning 'All Earth'). The climate was predominantly hot and dry, because of the close proximity to the Equator. Scotland was landlocked, forming part of the much larger continent.

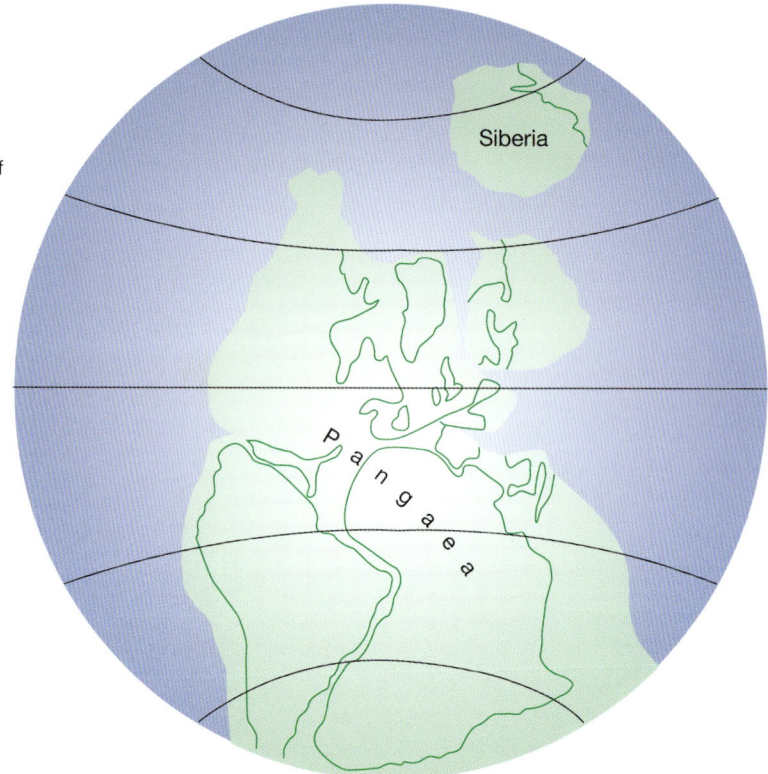

The new geography that emerged at the beginning of the Jurassic Period, some 200 million years ago, was very different from earlier times. America had started to drift westwards, powered by undercurrents in the Earth's mantle, and much of the western and southern areas of Scotland lay under a warm shallow sea. The precise location of the coastline is not possible to pin down as the evidence has been removed by subsequent erosion, but sufficient clues still exist to estimate its likely position.

from the nearby mountains and dumped their burden on the flatter ground. Exotic reptiles dating from these times are preserved in the rocks around Lossiemouth on the Moray coast, but no such spectacular finds have been recovered to date from the Triassic strata on Skye and Raasay.

Massive changes occurred at the beginning of the Jurassic Period in terms of the arrangements of the continents. Pangaea started to disintegrate as the Atlantic Ocean began to open. Europe was ripped from America as a consequence of these continental rearrangements. There was also a huge rise in sea level that left much of the land that was to become Scotland under a shallow tropical sea that teemed with life. Evidence for some of these environmental changes is to be found on the Trotternish Peninsula, particularly where thick sequences of sands, limestones and muds of Jurassic age are exposed along the dramatic coastal sea cliffs and foreshores.

Shallow-water conditions prevailed for most of this period and these early Jurassic sediments were deposited under tropical seas and in estuaries, deltas, mudflats and lagoons, which ranged from freshwater to brackish. The climate was tropical and coal-forming coastal swamps developed from time to time.

Remains of fossil plants, including conifer fragments and ferns, have been recovered from the Jurassic rocks of Berreraig Bay. The delicate structures of the plants have been fossilised by a process known as permineralisation. This means that the original plant structure decayed quickly to be replaced by an internal 'cast' of iron and carbonate-rich minerals that faithfully mimicked the original three-

Life in the Jurassic seas 165 million years ago. Ichthyosaurs are dolphin-like creatures that inhabited these waters. They had a streamlined body that was well adapted for hunting fast-moving prey and an elongated snout studded with razor-like teeth. They are rare in the fossil record in Scotland, but recent finds of bone material and teeth from the Jurassic rocks of Skye confirm their presence in these waters for a period of over 30 million years. Ammonites and belemnites are also shown in the foreground of this reconstruction.

dimensional structure of the plant. These rare finds add significant environmental information to the story. Recent research work suggests that these plant remains collected on the sea floor some 25–30km from the nearest land. But the plants originally grew on land, probably as part of a river delta, and were transported offshore as a raft of vegetation that later became waterlogged and then sank to the ocean floor.

Dinosaurs made an appearance in the fossil record of Skye in Jurassic times. An early clue was the discovery of dinosaur footprints at Staffin beach, stamped into the inter-tidal rocks. The footprints were made by passing dinosaurs probably foraging for food around the edge of the estuary that existed in this area during these times. The wet sands and muds carrying the imprints were later covered by sand and mud deposits. Over time, the whole sequence was turned to stone. Erosion by the sea has just revealed the rock layers carrying the footprints at the surface.

More recent dinosaur footprints found in the inter-tidal rocks at Valtos represent a considerable elaboration of the story. The individual footprints are the size of dustbin lids and represent ancient tracks made by sauropod dinosaurs as they plodded around the edge of their home patch – a lagoon fringing the Jurassic sea. The rocks that host these prints are of Middle Jurassic age, some 170 million years old. They are from a time when relatively few dinosaur fossils have been found worldwide. The trackways are detailed enough for them to be recognised as being made by distant relatives of brontosaurus and diplodocus. These amazing new finds will shed a brighter light on our understanding of some of the biggest animals ever to roam the planet.

This is the impression of a sauropod dinosaur footprint found in the inter-tidal rocks near Duntulm. The dinosaur must have plunged its foot into wet sediment which created an impression that erosion by the sea has recently excavated after remaining 'buried' in the rock for many millions of years. The little serrations around the footprint are impressions of the animal's claws.

Right. These are fossilised bones from the ichthyosaur *Dearcmhara shawcrossi* found in Bearreraig Bay. This marine reptile measured over 4 metres in length and lived in the Jurassic seas around 170 million years ago. The fossil remains were discovered by an amateur collector and later donated to The Hunterian, University of Glasgow. The name given to the fossil 'Dearcmhara' is inspired by the Gaelic term for 'marine lizard'.

Below. This is a reconstruction of the ichthyosaur *Dearcmhara shawcrossi* as it would have been in life.

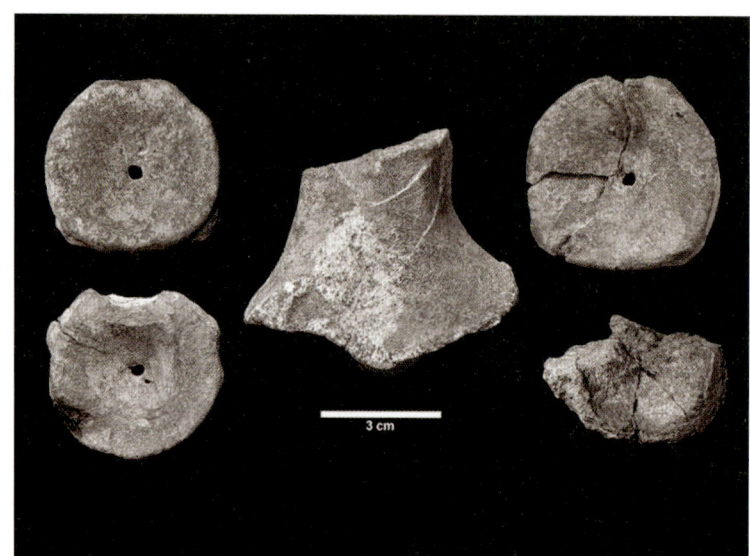

DINOSAUR ISLAND

Above left. Trackway of dinosaur footprints in rippled sandstone bed at Staffin Bay. The long, narrow impressions in the sand suggest that the creature responsible for the prints was a meat-eating dinosaur. The rippled sandstones are indicative of a lagoon-edge environment, where this predator would have lurked to catch prey or scavenge existing carcasses.

Above right. This 2cm-long serrated dagger-like tooth is from a theropod dinosaur. This group of animals was thought to be primarily meat-eating and this fearsome serrated edged-tooth would place this animal firmly in that category.

Left. This slab contains over 20 baby dinosaur footprints and an adult, all walking in the same direction.

Right. This is a reconstruction of a limb bone of a primitive sauropod dinosaur, recovered from Jurassic strata. The intact pieces of fossilised bone are dark and the light-coloured highlights are filler and glue that allows the original form of the bone to be revealed. This one bone is around a metre in length, giving an impression of how large the 'owner' must have been!

Below. Reconstruction of the plant-eating dinosaur *Cetiosaurus*, whose remains have been recovered from rocks on the Isle of Skye.

Left. At Valtos, the sandstones have been weathered by the elements into large cannonball-like structures that can be up to a metre in diameter. In other places the rocks 'weather-out' in an irregular manner to give a curious 'honeycomb' effect that can be seen in the sea cliffs near Elgol.

Below left. The cliffs at Elgol on the shores of Loch Scavaig provide an excellent section through rocks of Upper Jurassic age. They are just over 150 million years old and contain important assemblages of fossil ammonites. These strata have attracted the attentions of geologists since they were first studied over a century ago but have proved to be difficult to interpret as their original characteristics were significantly altered when the Skye volcano roared into life some 60 million years ago. The heat from the molten rocks of the Skye volcano (seen in the distance on the other side of Loch Scavaig) 'baked' the honey-coloured layers of sandstone that make up the sea cliffs at Elgol seen in this photograph.

Virtually all of the Jurassic rocks of Skye contain fossils. The most common are different kinds of oysters and other two-shelled organisms, gastropods (sea snails), ammonites (elaborate coiled shells of squid-like creatures), belemnites (like cuttlefish but with a hard, pointed internal shell), crinoids (sea lilies which, despite their name,

This is an array of typical fossils from the Jurassic strata of Skye; top is an ammonite. Middle left is a scallop; middle right is a belemnite; bottom left is a gastropod and bottom right is an oyster (also known as the Devils toenail!).

are actually animals related to sea urchins, and appear to grow from the sea bed like a plant) and corals. Fossils of microscopic organisms are abundant and are particularly helpful to geologists in sub-dividing rock successions.

At the end of the Jurassic Period and into the following Cretaceous Period, the sea level rose until most of the Hebridean area was covered by an extensive sea. Widespread deposits of organic oozes, rich in lime, built up on the sea floor, equivalent in age to the chalk found in southern and eastern England. But much of this blanket of sediments has been subsequently removed by erosion.

4
The Skye volcano

Around 60 million years ago, the North American and European continents continued to chart their divergent courses. A ferocious blast of hot rocks emanating from deep within the Earth's mantle, known as a mantle plume, drove this separation.

As the plates moved apart, the crust thinned and an extensive system of fractures developed that cut deep into the Earth's crust. Through these fractures, magma welled up and erupted in what was probably the most extensive volcanic episode ever experienced in north-west Europe. The products of these eruptions spread over both sides of the widening North Atlantic. Skye was the most northerly of these new eruptions in Scotland, with other volcanoes of the same age

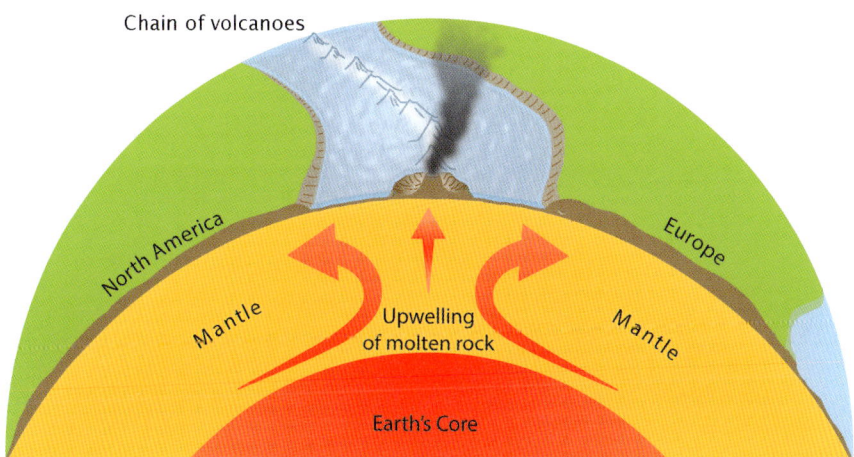

This cartoon tells the story of how the North Atlantic Ocean came to be. An upwelling of fiery molten rocks that came from deep within the Earth's mantle scorched the underside of the crust, heating it like a blowtorch. The continental crust cracked as a result and the two new plates moved apart, driven by the convection motion set up in the mantle. The Skye volcano, and associated erupting vents on Mull and Archnamurchan, were fed by this ready supply of molten rock. As these once active volcanoes moved away from the source of heat – the mantle plume – they have now fallen silent. But a chain of volcanoes that define the mid-line of the North Atlantic Ocean remain active to this day. The sporadic eruptions of basalt add new crust to the ocean floor, and by this continued process, the continents are forced further and further apart. Iceland sits over this ascending plume of hot rocks accounting for the volcanoes that are periodically active on the island.

These distinctive flat-topped hills are known as MacLeod's Tables. The hills are made from layer upon layer of basalt lava that erupted from the Skye volcano. The lavas were runny when erupted and each step in the hillside staircase represents a separate volcanic eruption. This is one of the 'Places to Visit' described in more detail on page 47.

active on St Kilda, Rum, Ardnamurchan, Mull, Arran and Ailsa Craig. Some 34 seamounts (mini-volcanoes) have been identified in this area that were former volcanic centres. Volcanoes were also active in Greenland, Baffin Island and the Labrador Sea area. This cacophony of activity accompanied the opening and further widening of the North Atlantic Ocean. The mantle plume continues to fire the volcanoes and geysers of Iceland to this day.

When it spluttered into life, the Skye volcano wasn't a short-lived affair. It was active, although not continuously, for almost 5 million years. The first evidence is of thick sequences of runny lavas that poured onto the land surface previously flooded by seas of Cretaceous age. The initial outpourings rapidly built up a vast lava plateau.

In northern and western Skye, the area is dominated by the stepped or 'trap' landscape created by lava flows. These lava flows are horizontal or gently sloping and form distinctive flat-topped hills with stepped sides. The well-known MacLeod's Tables in Duirinish are an excellent example.

Individual lava flows can be between 10 and 25 metres thick and are not necessarily uniform throughout. At the top and bottom they are usually broken and pitted with holes. These are former gas bubbles that date from the time the lavas were erupted. The central parts of the lava flows are usually more solid, some with distinctive columns, reminiscent of Fingal's Cave on Staffa and the Giant's Causeway in Northern Ireland.

THE SKYE VOLCANO

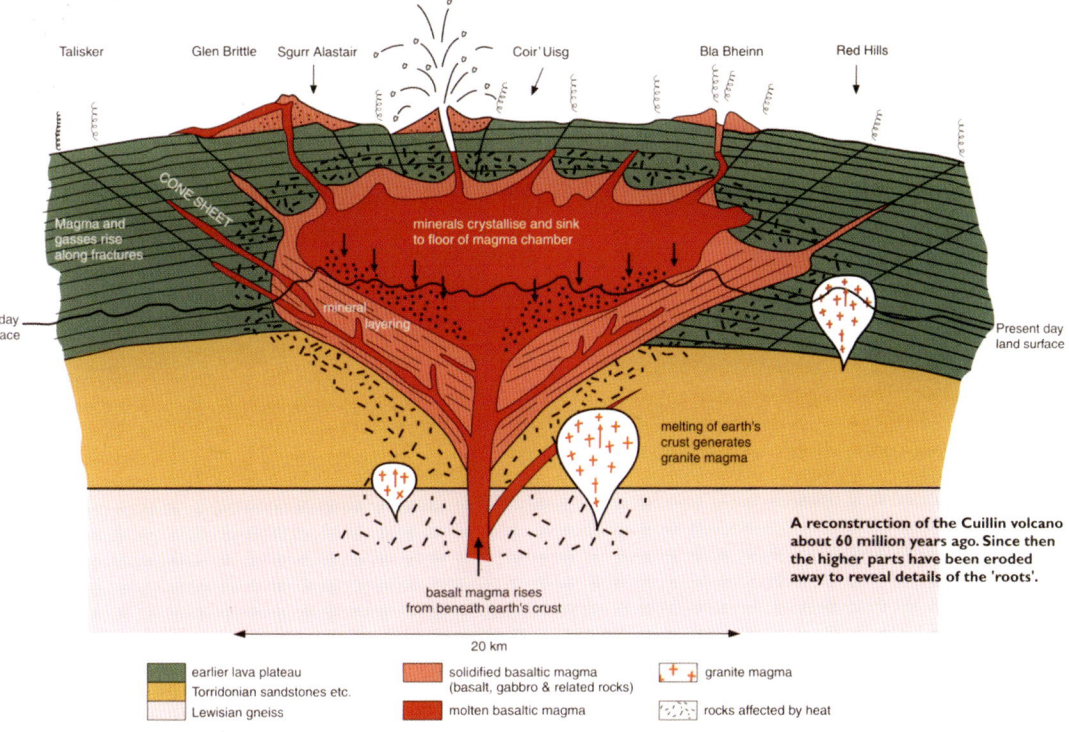

A reconstruction of the Cuillin volcano about 60 million years ago. Since then the higher parts have been eroded away to reveal details of the 'roots'.

This is a slice through the Skye volcano as it would have looked 55 million years ago. The thick pile of lavas accumulated layer upon layer, with each new burst of volcanic activity adding to what had gone before. The molten core of the volcano was primarily base-rich magma that much later gave rise to the Black Cuillin. The predominant rock type is known as gabbro. Magma of granite composition was generated when the hot rocks confined within the magma chamber came in contact with the surrounding older rocks. These rocks melted to create a series of pockets of molten rock with a granite composition. These balloon-shaped masses of granite were less dense than the surrounding rocks, so rose towards the surface like air bubbles rising through treacle. Over the next 50 million years, ice, wind and water planed away the overlying rocks and the inner workings of the volcano were exposed at the surface.

Accumulations of sedimentary rocks between the lava flows indicate that river systems and lakes developed between volcanic eruptions, and plant remains within the sediment suggest a sub-tropical or Mediterranean-type climate. Where a long time elapsed between successive flows, intense weathering formed red soils rich in clay that are similar to those developing today in wet tropical or sub-tropical climates.

Great reservoirs of molten rock accumulated at depth under the

THE SKYE VOLCANO

Opposite. This is a shot of a contemporary lava flow from Hawaii. The top surface of the flow has already started to cool forming a thin skin of solidifying rock. It would have been a similar view some 60 million years ago when the Skye volcano was active. Like the lavas of Hawaii, the molten rocks erupted on Skye were largely varieties of basalt that flowed easily across the land surface. They too would have glowed red hot until they cooled to form the latest addition to the geology of the Skye.

Left. This is a fossilised leaf from either the beech or hazel family that was recovered from sediments accumulated between lava flows in Glen Osdale. The eruptions of lava were separated by long periods during which time soils built up on the top surface of the most recent lava flow. The presence of leaves and other plant material provide good evidence for a verdant covering of vegetation before the landscape was once again scarified by the eruption of the next lava flow. These layers of sediment preserved between lava flows are like botanical time capsules, preserving evidence of the plant species that inhabited the landscape around 60 million years ago.

pile of lava. This magma chamber, which 'froze' when the volcano went silent, was exposed at the surface in recent geological times when ice, wind and water tore away the upper layers to reveal the volcano's internal plumbing and structures.

The cracks in the Earth's surface that fed these volcanic eruptions continued to act as pathways for molten magma long after the surface volcanic activity had ceased. Magma hardened in these fissures as vertical sheets of rock known as dykes. There are many examples of dykes on Skye, particularly on the coast where they cut through the sedimentary rocks and stand proud like walls because they erode less easily than the softer rocks around them.

In northern Skye, the magma also forced its way sideways to form

Right. The red layer represents a fossil soil that supported a variety of plant species, some of which are preserved as fossils – see page 29. There were sometimes long periods between eruptions when soils developed and vegetation flourished. The upper layer in the photograph is a lava flow that swept across the early landscape extinguishing any plant life that may have grown since the previous volcanic eruption took place.

Opposite top. The Kilt Rock is a well-known tourist attraction on Skye. See 'Places to Visit' on page 43 for the site description.

Opposite bottom. This is a scenic view of Skye's Black Cuillin. Its serrated ridge of high peaks is a favourite haunt of climbers who pit their skills against some of the hardest pitches in Scotland.

sheets between the layers of sedimentary rocks beneath the lavas. These gently sloping sheets can be up to 90 metres thick and are called sills. Many display well-developed columns throughout most of their thickness. This dramatic effect can be seen in the aptly named Kilt Rock south of Staffin.

The Cuillin of Skye is acclaimed as the most spectacular mountain range in the British Isles. The arc of jagged peaks, almost 1,000 metres high, which makes up the main Cuillin Ridge, together with Sgurr na Stri and Bla Bheinn, are a delight for all visitors, particularly mountaineers.

Less spectacular but equally attractive in their own way are the Red Hills extending from Glamaig to Beinn na Caillich. The granite of these hills is more uniformly worn by the elements, so the hills have smooth outlines with few steep rocky faces.

The mountainous areas of today are the solidified remains of the magma chambers that fed the volcanoes. Because they cooled slowly, deep in the Earth, they now consist of a wide variety of spectacular and beautiful coarse-grained rocks. In the Cuillin these rocks are mostly gabbro – a coarse-grained equivalent of the basalts that make up most of the lava flows that can be seen on Skye.

THE SKYE VOLCANO

Above. The Red Cuillin are the excavated roots of the Skye volcano. The granite formed deep in the crust and was exposed at the surface as kilometres of rock were planed away by ice, wind and water. The final moulding took place during the ice age when the hills, such as those above, were rounded into a smooth and regular outline by the passage of ice.

Opposite top. Preshal Beg in south-west Skye is a pond of basalt lava that accumulated on a surface of previously erupted lavas.

Scattered around both the Cuillin and the Red Hills are rocks which themselves consist of a jumble of rock fragments of various types, ranging from fine dust particles to huge blocks. These outcrops are the remains of volcanic 'vents' – the pipes through which the magma chambers were connected to the land surface, allowing highly explosive eruptions to take place. The vents contain materials brought up from deeper levels as well as the products of the eruption, such as lava and ash, which have collapsed back down the vent. Good examples of this phenomenon can be seen in the Kilchrist – Kilbride area between Broadford and Torrin.

A distinctive feature of the gabbros in some parts of the Cuillin is a very obvious layering reminiscent of that seen in many sedimentary rocks. On closer inspection this is seen to be due to a repetitive variation in the proportions of the individual minerals that make up the rock, such as feldspar, olivine and pyroxene. This happened during the cooling of the magma, as the individual minerals crystallised in a definite order and either floated or sank according to their density relative to the remaining magma. In general the layering dips inwards towards the centre of the magma chamber which, as a result, has an internal structure resembling a stack of saucers.

THE SKYE VOLCANO

This slab of rock, rendered smooth by the passage of the ice, is a slice through the magma chamber of the Skye volcano. Layers of minerals settled on the floor of the magma chamber just as sands or mud might accumulate on the lower reaches of an ocean floor. The diagram on page 27 shows the magma chamber and how, as new crystals form, they drop to the bottom forming a build-up of layers. This photo shows some of the layers that formed in this way. Minerals crystalised and settled out in response to the changing chemistry of the magma and these complex relationships have been the subject of detailed study for a century or more.

5
The Ice Age

For most of the next 50 million years, 'Scotland' basked in a subtropical climate. But its trajectory was ever-northwards into colder climes. By the beginning of Pleistocene times some 2.4 million years ago, natural variations in the amount of sunlight reaching the Earth's surface caused the climate to cool dramatically. The Ice Age had begun.

The changes in climate became even more extreme about 450,000 years ago, since when there have been four long, intensely cold, glacial episodes separated by short warm inter-glacials at roughly 100,000 year intervals. Surface layers were swept away by ice in the first major widespread glaciation. It is after this event that Skye probably first became an island, albeit initially entirely swathed in ice.

During the last widespread glaciation, which occurred between 30,000 and 14,700 years ago, Scotland was covered by an ice sheet. It enveloped all but the highest pinnacles on Skye and swept away most of the older glacial deposits. Following a sudden climatic warming about 14,700 years ago, the ice cap started to melt. The freshly exposed

Ice has polished the rock surface to create an impressive glacial pavement in Coire Lagan.

THE ICE AGE

Core samples have been taken from more than 20 peat bogs in Skye. The painstaking identification of pollen grains and insect remains layer by layer in the peat reveal how vegetation, and hence climate, have changed since the ice melted. Radiocarbon dates on pieces of wood and peat reveal when the changes occurred.

stony soils were soon colonised by pioneer vegetation such as grasses, sedges, club-mosses, 'alpine' herbs, and dwarf willow. Next came a mosaic of juniper scrub, crowberry, heather, grassland and birchwood.

Summer temperatures similar to those of today occurred until about 13,500 years ago, when the climate cooled again. By 12,500 years ago, a tundra environment had returned and arctic conditions prevailed.

During a final, relatively short, glaciation 12,500 to 11,700 years ago, corrie glaciers and substantial areas of ice and snow accumulated in the mountains. The surrounding lowland areas were frozen wastes, similar to the conditions found today in parts of arctic Europe and Asia.

Some 11,700 years ago, the climate warmed rapidly and the glaciers melted. The initial pioneer vegetation was soon replaced by birch and hazel woodland, ferns, tall herbs, and scattered willows. Oak and elm arrived on Skye about 10,000 years ago, together with some pine. Alder arrived much later. Mixed birch–hazel–oak woodland flourished until the climate became wetter some 6,000 years ago. At this time, people first began felling trees to plant crops – the beginning of human influence on the landscape.

Grass and heather moorland began to expand at the expense of the woodland and blanket bog started to accumulate. Pine stumps, regularly found in the peat, are proof of a short-lived expansion of 'Caledonian' pinewood about 4,500 years ago.

Approaching modern times, large areas of woodland survived in the south of the island until about 300 years ago, when cattle grazing became more widespread. As the trees were felled, the human factor became still more telling with increasing use of wood as fuel.

6
Dynamic landscapes

The landslides of northern Skye are unrivalled in Scotland. The best examples fringe the great escarpment of Trotternish where a thick pile of basalt lavas rests on relatively weak sedimentary rocks of Jurassic age.

With the march of time, the sedimentary rocks gave way to the great weight of the lavas, resulting in enormous landslides, the creation of awesome labyrinths of huge blocks and pinnacles bearing the evocative names of the Quiraing, Table, Needle, Prison, Dùn Dubh and the Old Man of Storr.

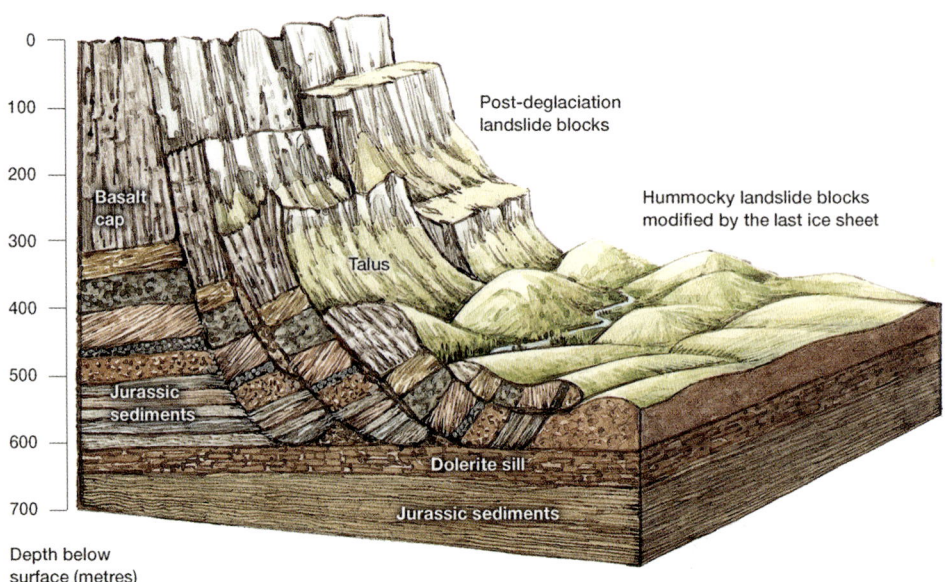

The landslips of the Old Man of Storr and Quiraing are impressive features of the Trotternish landscape. They formed largely after the ice melted. The pile of Jurassic sediments and later lavas that built the Trotternish escarpment just simply collapsed under its own weight. A great jumble of blocks of rock cruised downslope forming an impressive labyrinth of pinnacles and disconnected lumps of rock. It is certainly a dynamic landscape that is still on the move, but ever so slowly. It hasn't reached its final destination yet.

DYNAMIC LANDSCAPES

The largest and most recent features formed during the last 15,000 years. An earlier generation of landslide deposits, occurring further away from the escarpment, must have moved much earlier because they have been smoothed by the action of glaciers and are capped by sands and gravels left by the ice. Although the Trotternish landslides are the most famous, there are others along the coast of northern Skye and in Glen Uig.

Left. Quiraing landslip – more land on the move. It is part of the same landslip at the Storr and together they constitute one of the largest 'mass movement' features (also known as a landslip) in Great Britain. As with the Storr, some of the slippage took place before the ice age started. But the majority of the mass movement happened probably immediately after the ice melted at a time when the cliffs, steepened by glacial erosion, could not support its own weight following the withdrawal of the supporting buttress of the glacial ice.

Below. This view of the Storr landslip gives an impression of the extent of this feature. The land on the move extends from the basalt escarpment almost to the sea.

7
Living landscapes

This expanse of bogland has been exploited in recent times by peat cutting. Traditional tools are used to harvest the peats. After they been stacked and dried, the peats are used to heat the homes of some local residents.

This dynamic and diverse geological legacy has given rise to a varied landscape today. The land is predominantly mountainous, dramatically so in places, with a mosaic of natural habitats occupying the ground from shore to mountain-top. The geology has had a key role in shaping how the landscape looks today. The Cuillin, with peaks of jagged gabbro and rounded granite hills, are bare. They were sculpted and polished by the rasping sandpaper-effect of the ice into the familiar shapes we recognise today. Some of the higher peaks of the Black Cuillin stood proud of the last ice sheet, so their edges are serrated and sharp. The rounded profile of Marsco in Glen Sligachan is typical of the granite peaks that felt the full force of the glaciers.

Many hectares of the lower areas are occupied by peatland, little disturbed by human activity over the millennia. Along the broad valley-bottoms excavated by the ice, areas of blanket bog, dominated by Sphagnum moss habitat, are widespread.

Further up the mountainside, small patches of ancient and semi-natural upland woodland thrive. Birch, hazel and oak are the dominant tree species. Flowering plants are also present. Arctic mouse-ear, rock whitlow grass, alpine meadow grass and mountain avens are all rarities that have been recorded on Skye.

The coast provided some unusual habitats. The dazzling creamy-white beaches known locally as coral sands are created by the

Small patches of native woodland are common in the lower ground of the straths and glens of Skye.

Right. This delicate flowering plant – *Dryas octapetala* – has survived on the lime-rich soils around Torrin since the last glaciation.

Below. The 'coral sands' of Dunvegan are a local visitor attraction. They are a dazzling white colour and consist of broken fragments of the lime-rich algae *Lithothamnion* that grows in the offshore waters. Fragments are washed ashore regularly that replenish the beach.

calcareous alga *Lithothamnion*, which has a remarkable resemblance to branching coral. These beaches are also known as the 'singing sands' because of the distinctive sound they make underfoot. The algae flourish in sheltered bays around the west coast. Broken fragments are continuously washed up, and these unusual sands were used locally to 'lime' acid, peaty soils.

Machair is a coastal habitat with a strong cultural land-management tradition. It derives from shell sand accumulated offshore, then broken by the waves and driven inland by wind and tides. Crofting on the machair has a strong tradition, particularly in the area around Camasunary on the shore of Loch Scavaig.

The geology of Skye is the friend of the magnificent golden eagle, as the many craggy cliffs provide ideal ledges for nesting sites. This icon of Scotland has maintained a sustainable population on the island for a number of years now. Its beauty and grace are the stuff of Scottish legend and it will live long in the memory of anyone who is lucky enough to see this surprisingly large bird slowly glide its languid and unhurried way across the sky.

Up close and personal with a golden eagle.

8
Places to visit

The places of geological interest to visit on Skye are many and varied. What follows are just a few suggestions. Where you go very much depends on your particular interests and the time you have available. The OS Landranger Map series numbers 23, 32 and 33 will help you to navigate between these sites.

Map showing the location of all the places to visit.

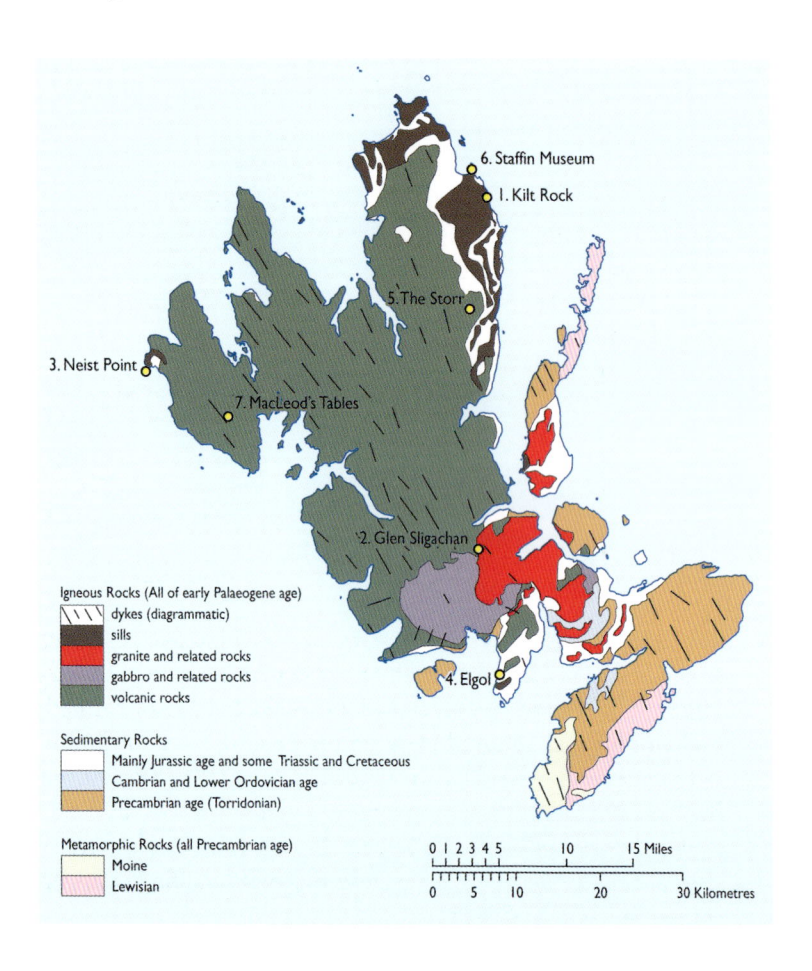

Kilt Rock.

1. **Kilt Rock** – this is a very visitor-friendly spot! There is a good-sized car park with plenty of interpretative assistance to help visitors understand the local geology. A thick sheet of formerly molten rock was squirted between pre-existing layers of Jurassic sediments whilst the Skye volcano was active. The relationship between the two very different types of rock is clearly displayed in the cliff sections and can be seen from a viewing platform. A stream tumbles over the cliff-edge at this point, adding considerably to the drama of the place.

2. **Glen Sligachan** – this glacially carved glen runs through the heart of the Skye Cuillin. A footpath runs south from the Sligachan Hotel to the sea. The more adventurous will want to complete a 20km loop that takes in Loch Coruisk and the infamous 'Bad Step'. But, for the less ambitious, the path offers the possibility of a gentle stroll, providing views of some of the most spectacular geology on the island. To the east of the footpath lie the Red Cuillin – rounded granite hills including the distinctive outline of Marsco. To the west lie the serrated peaks of Bla Bheinn and many of the other peaks of the Black Cuillin.

Glen Sligachan.

PLACES TO VISIT

Neist Point.

3. **Neist Point** – this is the most westerly point on Skye. The road from Glendale takes the visitor most of the way and then there is a tarmacadam path that leads to the lighthouse. This structure, which is over 20m in height, commands the waters of the Minch. The cliffs below are made from cooled magma that invaded the adjacent Jurassic sediments, also known as a sill, similar in style to that seen at the Kilt Rock.

The beach at Elgol.

4. **Elgol** – (NG 520165) a thick succession of Jurassic strata are to be found on the foreshore at Elgol on the eastern shore of Loch Scavaig. The honeycomb weathering of the sandstone is a notable feature of the area, as are the dramatic views of the Skye Cuillin from this point.

5. **The Storr** – the origin of the Storr and Quiraing landslips has been described on pages 36 and 37. They can be appreciated at a distance or up close. The main tower, known as the Old Man, is over 50m high and a hugely impressive sight. The views eastwards across the water to the Isle of Raasay are equally worth the climb. To visit the Storr landslip and the Old Man, leave your vehicle in the car park at NG 509529 and proceed along the well-maintained path to the landslip.

PLACES TO VISIT

6. **Staffin Museum** – (NG 506657 – 25km north of Portree) – the museum in the small settlement of Ellishadder houses a range of exhibits that interpret the local geology. Some dinosaur footprints, including the world's smallest dinosaur footprint, are on show here, along with a reconstructed dinosaur leg bone and many other fossils from the Jurassic of Skye.
7. **MacLeod's Tables** – the lavas that flowed from the Skye volcano are a prominent landscape feature of the northern part of the island. There are myths and legends associated with Healaval Mhor and Healaval Bheag – the flat-topped hills known as MacLeod's Tables. It is said that Alasdair Crotach, seventh chief of the clan McLeod, brought a group of lowland nobles to Skye in the sixteenth century. In an effort to impress the Scottish nobility, he laid on a lavish feast on the flat summit of Healaval Mhor. He said of his open-air venue 'Have you ever seen a hall so spacious, a roof so lofty, a table so ample and a candelabra so ornate?' It's worth a visit – with or without a picnic.

Staffin Museum.

Acknowledgements and Picture Credits

Thanks are due to Professor Stuart K. Monro OBE and Moira McKirdy MBE for their comments and suggestions on the various drafts of this book. Dr Neil Clark also provided useful comments and images. I also thank Hugh Andrew, Andrew Simmons and Mairi Sutherland from Birlinn for their support and direction. Mark Blackadder's book design was up to his usual high standard. Scottish Natural Heritage, in association with the British Geological Survey, published the 'Landscape Fashioned by Geology' series that was the precursor to the new 'Landscapes in Stone' titles. I thank them for permission to use some of the original artwork and photographs in this book. David Stephenson and Jon Meritt wrote the original text for *Skye – A Landscape Fashioned by Geology*.

Picture credits

Title page Martin M303; 6 Sergejus Lamanos; 10 drawn by Jim Lewis; 11 drawn by Robert Nelmes; 12 drawn by Jim Lewis; 13 drawn by Jim Lewis; 14 BGS; 15 BGS; 16 drawn by Jim Lewis; 17 drawn by Jim Lewis; 19 Neil Clark; 20 Neil Clark; 21 Neil Clark; 22 (right) Neil Clark; (lower) Craig Ellery; 23 (both) I Sarjeant; J.G. Hudson; 24 BGS; 25 drawn by Jim Lewis; 26 Gary Smith/Alamy Stock Photo; 27 SNH; 28 Denis Budkov; 29 BGS; 30 Alan McKirdy; 31 (upper) Catalina Panait, (lower) SNH; 32 SNH; 33 (both) SNH; 34 SNH; 35 J.W. Merritt; 36 drawn by Robert Nelmes; 37 (upper) SNH, (lower) Martin MO3; 38 Daan Kloeg; 39 nagelstock.com/Alamy Stock Photo; 40 (upper) L. Gill/SNH, (lower) Bill Spiers; 41 Nadezda Murmakova; 42 drawn by Jim Lewis; 43 Karolina Grabara; 44 Martin Fowler; 45 Martin M303; 46 Sara Winter; 47 Neil Clark